OUR CHINA PROSPECTS

Memoirs of the
AMERICAN PHILOSOPHICAL SOCIETY
Held at Philadelphia
For Promoting Useful Knowledge
Volume 121

OUR CHINA PROSPECTS

Symposium on
Chinese-American Relations
at the
Autumn General Meeting
of the
American Philosophical Society
November 12, 1976

Edited by
JOHN K. FAIRBANK

The American Philosophical Society
INDEPENDENCE SQUARE, PHILADELPHIA
1977

The publication of this book was aided by
the John Louis Haney Publication Fund
of the American Philosophical Society

Library of Congress Catalog Card Number 77-79208
International Standard Book Number 0-87169-121-3
US ISSN 0065-9738

FOREWORD

Post-Mao China is gradually but inexorably becoming a world power, yet the United States still remains involved in the Chinese civil war out of which the People's Republic of China emerged in 1949. By our mutual security treaty of 1954 we are still allied with the defeated Nationalist Government of the Republic of China on Taiwan. Why this anachronism? When and how may it end?

The People's Republic seeks to be self-sufficient, yet it needs to import foreign technology. What are the prospects for American trade and technological exchange with China?

How may Japan and the Soviet Union impinge upon Sino-American relations? Are we heading toward stability or trouble in the Western Pacific?

As part of its observance of our Bicentennial, the American Philosophical Society at its Autumn General Meeting of November 11–12, 1976, featured a symposium on Chinese-American Relations, of which this book is the product.

Two of the contributors, the first and the last, have had long experience in the study of Chinese affairs since the 1920's. John K. Fairbank, Francis Lee Higginson Professor of History at Harvard University, first went to China as a graduate student in 1932 and has been intimately involved in the growth of Chinese studies ever since. O. Edmund Clubb, also an historian, served as a consular officer in China from the 1920's to the 1950's, first reporting on the Chinese Communist movement when at Hankow in 1932. Subsequently he represented the United States in a variety of key spots—at Vladivostok among the Russians, at Hanoi when it was taken over by the Japanese, in Central Asia at Tihwa (Urumchi) during World War II, and in Peking when it was taken over by the Chinese Communists. He concluded this distinguished career as director of the Office of Chinese Affairs, Department of State.

The other three contributors are younger men who have emerged as leaders in their fields. Akira Iriye, whose father, Iriye Keishiro, has been a pioneer historian of international relations in Japan, began his education in Tokyo, and completed it at Haverford College and Harvard University. He is now professor of international history at the University of Chicago and also chairman of the American Historical Association's Committee on American-East Asian Relations.

Dwight H. Perkins, professor of China Studies and Economics at Harvard, has been widely involved both in studies of Chinese economic history and in the analysis and planning of development in various countries of East Asia. He is currently chairman of Harvard's Economics Department.

Frank Press, chairman of the Department of Earth and Planetary Sciences at the Massachusetts Institute of Technology, is a leader in the study of earthquakes and headed the earth sciences delegation to the People's Republic. Since 1974 he has been chairman of the Committee on Scholarly Communication with the People's Republic of China, a central group which represents the American Council of Learned Societies, the Social Science Research Council, and the National Academy of Sciences in the conduct of exchanges with China. Dr. Press is now President Carter's science adviser.

The wide variety of approaches as well as the special skills manifest in these contributions has made it seem desirable to publish them for a wider public.

JOHN K. FAIRBANK

Contents

American Intervention and the Chinese Revolution

CHINA'S GREAT REVOLUTION has been accelerating for a century and is far from over. American intervention in Chinese life helped to precipitate the great revolution; but we have still not come to terms with it, nor it with us.

1. *Characterizing the Problem*

From the 1830's to the 1940's Americans in China, particularly our Protestant missionaries, imported into China such revolutionary Western values as individualism, science, legal rights, constitutionalism, nationalism, and warfare-in-the-service-of-nationalism. As the ancient Chinese culture and civilization gradually crumbled, we offered our models of how to reshape it in a "modern" (meaning Western) form. But these models—whether the Protestant social gospel after 1900, or parliamentary government in Peking after 1911, or businessmen's government later in Shanghai, or farm extension work out of Nanking after 1928, or the coeducational Christian colleges in the 1930's—all failed to reach the great mass of the Chinese people in the million villages where they lived. The great revolution when it finally came followed other models better suited to the task of mass mobilization in the cause of national industrialization and social change.

Did we applaud the eventual success of these native Chinese efforts to meet China's problems? No, we found them ideologically menacing and opposed them as evil. We fought the People's Republic of China (quite unnecessarily) in Korea. We kept it out of the United Nations till 1971. We are still allied militarily with its enemy the Nationalist Government of the Republic of China on Taiwan.

Meanwhile the People's Republic is headed in a very different direction. Since the Americans in China were foreigners representing the old establishment which many of them had helped to build up (in the form of business firms, colleges, hospitals, churches, and the like), the revolution drove them all out. The People's Republic of today is opposed to freedom of the press and of corporate enterprise, and to due process of law that protects both the individual and the corporation. China is also opposed to the stationing of troops in other countries, to arms sales abroad, to the rights of alleged criminals, to open government, the legal profession, drugs, advertising, stock markets, private automobiles, super-highways, and other salient aspects of American society. Not only do they not have these things; they say they do not *want* them. Imagine a life without automobiles or lawyers! This is what we call the Sino-American cultural gap.

Still another gap confronts us. We face in China a complex of historical issues that increasingly demand to be understood by American policy-makers. But the China field in the United States still suffers from an intellectual gap between history and the policy sciences. Almost no one dealing with current policy has a firm grasp of the attitudes, institutions, and style of China's traditional polity from 221 B.C. to 1911. What were the Chinese imperial doctrine and practice concerning territorial expansion? concerning the defense of the realm by sending expeditions beyond the borders? concerning maritime trade and colonies overseas? Political science does not grapple with these traditions. Instead it puzzles over current behavior in Peking, bemused by the way in which Maoist thinking mixes its Marxist lingo with Chinese literary allusions. Historians are no help either. To study the Ming dynasty is a life work. Why add Mao to the mixture? In short, the China field is too young in the United States and too little developed to permit a proper marriage of history and policy such as we normally expect in the case of Europe. I suppose an American high official who had studied Metternich's diplomacy would find the Chinese equivalent in a study of the tribute system. Though Mr. Kissinger when at Harvard was in the Government Department and never took our course, he had the historian's genius to perceive that a Chinese emperor could make peace with a barbarian chieftain only if that chieftain came to Peking as Mr. Nixon did in 1972.

Today, both the Chinese revolution and the American intervention (in Taiwan as part of China) are still under way, on opposite sides of the cultural gap—a situation that may unexpectedly blow up in our faces if we cannot make history the handmaiden of foreign policy.

Let us begin with the contrasts. China is the oldest continuous polity, a hold-over until 1911 from the age of universal kingship in river-valley civilizations like Egypt and Mesopotamia. Imperial China was contemporary at its birth with the empire of Alexander the Great, yet it survived into the twentieth century.

The United States is the most recent addition to the major Western powers, with a brief history that barely goes back to the late Ming, and dates its independence only from the reign of Ch'ien-lung—consequently lacking in historical perspective.

Yet in contrast, among major states today the Chinese revolutionary polity is the youngest, the American the oldest. We face vastly different problems—in China, how to industrialize a new nation composed 75 per cent of peasants and carry through a social revolution to wipe out the very ancient ruling-class tradition. In the United States, how to deal with the flux of a post-industrial technological civilization whose problems even defy listing.

Within these contrasts, let us seek order by envisioning two cognate traditions: (1) the tradition of continental bureaucratic administration and (2) the tradition of maritime trade and warfare. These traditions are of course universal ingredients in the history of major states; but in China and America they appear in inverse proportions. China with its million peasant villages and hordes of tax-gatherers has been predominantly *agrarian-bureaucratic*, with the military and the merchant class generally kept subordinate to the civil magistrates, who represented the emperor as the center of moral authority and social order.

The United States on the other hand is the latest in the long succession of *trading and fighting* societies that first appeared in the city-states of Greece and the Eastern Mediterranean, later in the Italian cities of the Renaissance, and eventually in the nation states of Western Europe that overran the world. Our United States comes out of a history as old as China's but even more diverse, full of seafaring, exploration, commerce, piracy, enterprise, seapower, colonization, conquest, imperialism, economic growth, and warfare. Where the Chinese state has had domestic rebellions, we have had foreign wars, averaging one per generation, great stimuli to nationalism and the arts—for example, to Chinese studies, which were created in this country by World War II (fought against Japan in China), by the Korean War (against China in Korea), and by the Ford Foundation, which recognized a problem, but not in time to keep us out of Vietnam.

I suggest that the great Chinese revolution of today is primarily in the continental agrarian-bureaucratic tradition, while the American intervention in China has been in the maritime commercial-military tradition. The American intervention has been possible because it converges with a Chinese maritime tradition that has come down from early times though generally subordinated in the history books to the dominant tradition of continental empire.

2. *The Chinese Maritime Tradition*

In brief, during some four centuries in late medieval times under the Southern Sung, Yuan, and early Ming from about 1050 to about 1450, China's main economic area was the Lower Yangtze and her economic life saw an increasing maritime trade both along the Chinese coast and overseas to Southeast Asia and also to Japan. The Southern Sung in the twelfth century opened ports, constructed harbors, encouraged sea trade, taxed it, and developed a navy.[1] The Mon-

[1] Jung-pang Lo, "Maritime Commerce and its Relations to the Sung Navy," *Jour. Economic and Social History of the Orient* **12**,1 (1969): pp. 57–101.

gols, in setting up their Yuan dynasty, in the thirteenth century, took over the Sung navy and mounted naval expeditions twice against Japan, using thousands of vessels, and also to Vietnam and to Java.

This great age of Chinese seafaring, long before the era of European exploration and expansion overseas, was facilitated by China's preeminence in naval architecture: the sternpost rudder, transverse bulkheads, and the use of the compass, which were only the more obvious symbols of this long-established superiority.[2]

China's naval age came to a climax in the seven early Ming expeditions sent between 1405 and 1433 into the Indian Ocean. Using hundreds of vessels, some of them 300 feet long with half a dozen masts, and rudders (which have been excavated) as big as 20 feet across, these armadas carried as many as 28,000 men including cavalry. They visited some thirty countries all the way to Hormuz in the Persian Gulf, Aden, and the east African coast. They bestowed presents, solicited tribute, fought battles at least in Sumatra and Ceylon, and brought numerous potentates and tribute envoys to see the Son of Heaven at Peking.[3] For a generation China was the naval power of Asia, a century before the Portuguese sailed around Africa.

Yet this Chinese age of seapower came to an abrupt end after 1433, for reasons not yet fully studied. Strategically, the Ming Chinese had to concentrate on fighting off the Mongols from Inner Asia. The court felt no need for overseas trade or colonies. Merchants could not set policy. As the population grew in early modern times, China centered more in her interior provinces. Finally, the continental agrarian-bureaucratic tradition was reasserted by the Manchu conquest of the seventeenth century, which for a time even closed the seacoast to foreign contact and eventually confined foreign trade largely to Amoy and Canton.[4]

[2] Joseph Needham, *et al.*, *Science and Civilisation in China,* vol. 4, part III, *Civil Engineering and Nautics* (Cambridge, Cambridge University Press, 1971), pp. 379–699.

[3] For the fullest account see J.V.G. Mills, transl. and ed., *Ma Huan, Ying-hai sheng-lan "The Overall Survey of the Ocean's Shores"* (Cambridge, Cambridge University Press, 1970).

[4] Jung-pang Lo, "The Decline of the Early Ming Navy," *Oriens Extremus* **5**, 2 (1958): pp. 149–168.

By the time Europeans—Portuguese, Spanish, Dutch, British, and French—set up colonies in Southeast Asia in the sixteenth and seventeenth centuries maritime China was represented there only by the ubiquitous Chinese junk trade and the overseas Chinese merchants in all the port cities of Malaya and the Indies.

European sea trade with China got going through the channels of this Chinese junk trade, at times using its goods, its pilots, and its supercargoes, sometimes even its ships. Thus was born an alliance between maritime Europe expanding eastward and maritime China expanding southward though disesteemed by its home government.[5]

The contrast was at the top, in the government attitudes. The Ch'ing dynasty of the Manchus, oriented toward Inner Asia, forbade its subjects to go overseas and punished them if they returned, while the King of Portugal, which was one-hundredth the size of China, styled himself "Lord of the Conquest, Navigation and Commerce of Ethiopia, Arabia, Persia and India." While China's ruler was at the center of his civilization, Portugal's ruler was on an expanding frontier—a have-not and therefore more aggressive.

The annals of European colonialism in Southeast Asia do not stress the overseas Europeans' partnership with the overseas Chinese, but it emerges clearly from the record. Chinese merchants generally handled the retail trade of the region, Chinese artisans supplied skilled labor, Chinese middlemen functioned as officially licensed monopolists and tax-gatherers, in a thin stratum beneath the European rulers and above the native masses, as one element in the plural colonial society. Although the dynastic rulers in Peking gave them no help, Chinese merchants, migrants, and entrepreneurs in Southeast Asia found their place under the Europeans' wing as commercial and fiscal collaborators in colonialism over the local peoples. Maritime China thus had its colonial phase under European auspices in the Dutch East Indies, Malaya, the Philippines, and French Indo-China. This Chinese par-

[5] Two recent studies are by Ng Chin-Keong, "Gentry-Merchants and Peasant-Peddlers—The Response of the South Fukienese to the Offshore Trading Opportunities 1522–1566," *Nayang Univ. Jour.* **7** (1973): pp. 161–174; "The Fukienese Maritime Trade in the Second Half of the Ming Period—Government Policy and Elite Groups' Attitudes," *ibid.* **5** (1971): pp. 81–100.

ticipation was attested in the Chinatown sectors of Batavia (Jakarta), Malacca, Penang, Manila, Cholon, and Haiphong. The success of Raffles's founding of Singapore in 1819 was assured when the Chinese junk trade moved in to use the new port. Today Singapore is no longer a British colony but a Chinese republic under Prime Minister Lee Kuan-yew and a maritime Chinese leadership.

But the chief arena of maritime China's modern growth, still under the foreign wing, was in the opening of China to Western trade and contact. When the British after 1800 found that they could pay for teas from Canton by selling to China opium from India, they unwittingly began an unholy partnership with equally adventurous Chinese entrepreneurs who quickly developed China's domestic channels for the illicit distribution of Indian opium. British private merchants from India, licensed by the East India Company government there, brought the drug to the China coast. The Chinese did the rest. Recent work suggests that the Opium War, though it would no doubt have occurred in any case, was properly named, for the British-Indian opium trade supplied the *casus belli*, the strategy and rationale, many ships, pilots, and interpreters, and most of the ready cash to pay for the British expedition.[6] All of this activity required the collaboration of Chinese who were not yet modern patriots.

Out of the war came Hong Kong, a permanent unsinkable opium warehouse and haven for Chinese secret society dissidents. By the 1880's when Indian opium imports were losing out to Chinese domestic opium, Peking's inspector general of Customs, Robert Hart, was able to set up a Maritime Customs blockade of Hong Kong and reduce its free smuggling trade. But Hong Kong remained a haven for dissidents. It gave Sun Yat-sen both his education and his first base for a republican revolutionary effort in 1895 and later.

Similarly the treaty ports, under foreign consular administration during the unequal treaty period from 1842 to 1943, were havens for Chinese Christians, political refugees, compradors, and pioneer journalists. Modern China soon cen-

[6] Peter Ward Fay, *The Opium War 1840–1842* (Chapel Hill, University of North Carolina Press, 1975).

tered in Shanghai, a legal anomaly where Peking retained sovereignty but the Shanghai Municipal Council ran a modern city. In Shanghai Chinese merchants quickly learned the skills of foreign merchants and soon, as compradors under contract, were running the Chinese side of the foreign trade, selling off the Western imports, buying the Chinese teas and silks for export. Western merchants loved Shanghai life because Chinese not only did their housework but also did their business for them. Out of the early comprador experience came the modern Chinese businessmen, bankers, and industrialists of later generations, who today are major participants in the international trading world.

If we view Taiwan in this context, it is no accident but the fruit of a meeting of the maritime West with maritime China, with a half-century's assistance after 1895 from maritime Japan. All these maritime trading societies tend to pluralism, the acceptance of due process of law as superior to ethical teachings, the protection of private property and capitalist enterprise. Taiwan is part of the world we Americans live and struggle in. Our mutual security treaty of 1954 continues our American intervention against Taiwan's enemy, the great Chinese revolution on the mainland. We have inherited this intervention from a past century but it does not wither away. On the contrary it grows apace. Chase Manhattan Bank functions in Taipei and the island's tax-free processing zones assemble most of our radios and television sets.

3. *The Continental One-China Tradition*

Curiously the government of Taiwan has inherited from its founder Chiang Kai-shek the doctrine of One-China that has inspired all would-be unifiers of the Chinese realm and still constitutes the central political myth of the continental agrarian-bureaucratic state. I need not expatiate here on the vitality of the One-China ideal, since it has been enshrined for several years now as a basic article of United States-China policy. Suffice it to say that a doctrine formerly held dear by a thin ruling class has now been implanted, by the revolution, in the consciousness of 900 million Chinese patriots. One-China is not an idea we should casually denounce or try to counter.

One-China has also been stoutly maintained as an article of faith on Taiwan, and we can expect no help from there if we suggest that there is a touch of unreality about it. For years now no China specialist has uttered the taboo phrase, "Two Chinas," except perhaps in sleep, delirium, or drunkenness, when not in his or her right mind. The equation is that One-China plus One-China equals One-China. The One-China doctrine is a sacred tenet in Taipei not merely as an inheritance from Chiang Kai-shek but as an inheritance from two thousand years of Chinese history. Whoever has aspired to rule in China has usually found himself compelled to try to rule *all* China, not just part of China. Before we dismiss the One-China concept as an out-of-date holdover no longer valid in the present day, let us stop and consider its roots. How did One-China become the central myth and political ideal of the Chinese state?

First, if we look at China's historical experience: in the Chou period from about 1100 B.C. to 221 B.C., North China saw a proliferation of city-states, each one led by a ruling family in a walled town with its surrounding countryside. Two hundred or more such principalities grew up as the Chinese population increased and spread over the North China plain. Most of them were nominal vassals of the weak Chou king at his capital near present-day Sian. But the growth of so many states was soon accompanied by a process of concretion as the more powerful city-states absorbed others roundabout, until by the eighth century B.C. there were about ten major states. Finally in the Period of Warring States from 403 to 221 B.C. China experienced a political-military free-for-all in which the leading states fought and devoured one another until one of them eventually climbed on top and created the unified Chinese empire.

Since Confucius and the other philosophers lived in this era of turmoil, one of their chief concerns was how to achieve peace. The Confucian teaching of social order, which has since inspired about as many people as the teachings of Christianity, was born out of this long struggle against violence and disorder. Chinese contemplating the warring states of the international world today have a strong sense of *dejà vu*.

Next, the unified empire, once established, consolidated

its position not only by creating methods of military control and bureaucratic administration over its vast territory but also by developing an orthodoxy that would hold the social order together by getting each individual to have the correct attitudes and beliefs useful for that purpose. In other words, the Chinese empire learned how to remain unified by indoctrination as well as by force, administrative devices, and other means. The ideal of unity, personified in the Son of Heaven, was a chief factor making for unity.[7] One-China has been achieved in practice about two-thirds of the time since 221 B.C., quite enough to keep the idea intact and vigorous.

Finally, in the modern revolution the need for unity has been obvious as a means of saving China from foreign domination. Imperialist aggression in the 1890's in China inspired the rise of nationalism, and revolution has been the result. Revolutionary China today aims at the unification, regeneration, and development of the whole Chinese realm. On this point Sun Yat-sen, Chaing Kai-shek, and Mao Tse-tung have been in basic agreement. Chinese political thinking in Peking, as in Taipei, posits One-China.

We can begin to understand this nationalistic view by recalling the most-remembered part of our American national history, the War Between the States. (It was fought over the issue of unity, and the Union side won.)

But Americans are devoted to a sentimental ideal—self-determination for under-dogs at a distance. While denying the several states' right of secession at home, we are prone to urge plebiscites upon troubled areas abroad. That the local people should decide upon their own form of government appeals to our town-meeting tradition. However, the real issue in self-determination is the size of the unit to be self-determined. We are on record as favoring for ourselves a unit of continental size. So are the Chinese.

The best reason for our accepting the doctrine of One-China in the Chinese realm is that there are 800 or 900 million Chinese who believe in it or, if they disbelieve, have no alternative to propose. We can hardly make up their

[7] For a broad perspective on continental China's tradition see Ray Huang and Joseph Needham, "The Nature of Chinese Society: A Technical Interpretation," *East and West* (IsMEO, Rome), n.s., **24**, 3–4 (Sept.–Dec., 1974): pp. 381–401.

minds for them. The Republic of Singapore is not and never has been part of the Chinese continental realm. But Taiwan has been part of China for three hundred years, first as a prefecture of Fukien province in early modern times, then as a province of the Chinese empire from 1885 to 1895 when the Japanese took over.

I have now suggested how the American intervention in nineteenth-century China not only helped inspire the great revolution but also allied itself with the minor Chinese tradition of maritime China, the tradition of seafaring and overseas trade which naturally seeks opportunity for individual enterprise and corporate growth protected by sea power. This history lies behind our military and commercial alliance today with Taiwan.

I have also sketched briefly but I hope conclusively the central role in Chinese political life of the One-China doctrine that is now an article of faith in both Peking and Taipei. This history lies behind Peking's claim to sovereignty over Taiwan as part of China, a claim our present policy does not challenge.

We are thus left with a gap between theory and practice, between the acknowledged ideal of China's unity and the obvious reality of our dealing with two regimes, which are still parties to an unresolved civil war and maintain no contact except that of shells or propaganda leaflets fired in anger. Given our other problems, we tend to leave the Taiwan issue on a back burner, but it is a time-bomb nonetheless and carries within it the potentiality of another Chinese-American war.

Some Americans like to assume that, since we have got this far dealing with both Chinese parties, we might as well let things drift along further. Such optimists overlook the fact that our present China policy is based on the idea of normalization, meaning the withdrawal of our diplomatic recognition of the Republic of China. We should realize that the setting up by the People's Republic of a liaison mission in Washington in 1973 was a concession in the spirit of going at least halfway with us. Peking for twenty-seven years has refused to sit in international bodies or appear in foreign capitals where Taipei is formally represented. Yet Mao and

Chou were willing to send their ambassadors to Washington as a special concession to break the impasse in Sino-American relations, even though their Liaison Office ambassadors would be outranked by the ambassador of the Republic of China. This anomaly is now in its fourth year. The longer it goes on, the more humiliated and angry Peking may well feel at our pluralistic indifference to its second-class status in our capital city. Now that Mao and Chou are gone, other Chinese leaders may make political capital out of this, the Shanghai communiqué of 1972 may be called into question, and the Peking-Washington rapprochement may retrogress. But we hope not.

JOHN K. FAIRBANK
Francis Lee Higginson Professor of History,
Harvard University

Origins of the Sino-American Cold War

IN ORDER TO UNDERSTAND the period of intense hostility in Sino-American relations which lasted throughout the 'fifties and 'sixties, we must look at the two decades of the 1930's and the 1940's. Second, we must add Japan to the bilateral relationship because it is very difficult to divorce United States policy toward China from that toward Japan, and this was particularly the case during the 'thirties and 'forties.

1. *America and the Development of Asia*

We are familiar in a general way with the history of United States-Chinese-Japanese relations during those years. The United States and Japan clashed in China after Japan's invasion of Manchuria in 1931 and the rest of China in 1937, and this conflict eventuated in the war of 1941–1945. After Japan's defeat, however, it was China's turn to collide with the United States. Soon America and Japan began behaving as allies against the People's Republic of China in a cold war confrontation which in 1950 developed into actual war in the Korean peninsula.

In tracing this story of wars and policy reversals, most writers have supposed that China and Japan have stood for two choices open to American policy in Asia. The United

States, it has been argued, has either befriended Japan against China or supported China against Japan because to do otherwise might result in an impossible situation: a Sino-Japanese combination against America. Perhaps no writer has expressed this theme more clearly than Mr. George F. Kennan, who says in his *Memoirs* that his one single contribution to policy-making as a State Department official lay in his 1948 recommendation for a shift in United States policy in East Asia so that the United States would have Japan, not China, as a potential ally. According to him,

> Americans, laboring under that strange fascination that China has seemed to exert at all times on American opinion, tended to exaggerate China's real importance and underrate that of Japan. . . . We Americans could feel fairly secure in the presence of a truly friendly Japan and a nominally hostile China . . . but the dangers to our security of a nominally friendly China and a truly hostile Japan had already been demonstrated in the Pacific war. Worse still would be a hostile China *and* a hostile Japan.[1]

It was with these views in mind that Kennan went to Tokyo early in 1948 to confer with General Douglas MacArthur, supreme commander of the allied occupation forces. They came to a meeting of minds, and soon afterwards the National Security Council and President Harry S. Truman approved a policy to stress Japan's economic recovery, political stability, and dependence on American military support so that a new framework of United States-Japanese cooperation would establish a solid base for peace and security in Asia. The peace treaty of 1951 was a product of this policy.

While I do not wish to belittle Mr. Kennan's contributions to policy or to our understanding of the past, it does seem that his description of the shifts and turns in the United States-Chinese-Japanese triangle tends to mask another important theme which runs through the history of these triangular relations. That is the theme of continuity, a theme I would like to stress in this essay. To get at this point, one may profitably turn to the conversations Kennan had with General MacArthur in Tokyo. According to a report of his trip that Kennan wrote upon his return to Washington, Mac-

[1] George F. Kennan, *Memoirs 1925–1950* (Boston, Little, Brown, 1967), pp. 374–375.

Arthur summarized his views this way: "The great significance of this occupation lay in the fact that it was bringing to the Japanese people two great appreciations which they had never before possessed and which were destined to revolutionize their thinking, namely, democracy and Christianity." He went on to say that there were one billion Asians living on the shores of the Pacific, a fact that would ensure that "great events of the next thousand years would transpire in this area." The United States, declared MacArthur, "had the opportunity, through the Japanese, to plant the seeds of the appreciation of Christianity and democracy not only in Japan but throughout the whole enormous area and to bring to these billion people . . . the blessings of freedom and of a higher standard of living." If the Americans were to succeed in this mission, the general concluded, "we might fundamentally alter the course of world history."[2]

Now there are two interesting things about these statements. First, ideas similar to, even identical with, these views had been expressed about China. If we substitute "China" for "Japan" in MacArthur's remarks, we get a good example of what numerous Americans had been saying about China. In other words, in American perception China and Japan were often interchangeable. Second, despite this interchangeability, in 1948 it seemed to some like MacArthur that Japan and not China was likely to emerge as America's principal partner in Asia. These two facts tell a great deal about the history of United States-Asian relations in general and Sino-American confrontation in particular.

The interchangeability in American perception of China and Japan is related to the theme of continuity in United States-Asian relations. The above remarks by General MacArthur reflected certain idioms or modes of thought that had characterized the way Americans looked at Asia. It was believed, as he said, that East Asia's billion people would present a fundamental challenge to the United States because America was a Pacific as well as an Atlantic power and because those Asians could be either a threat to or an accommodating force for American ideals and interests. The best

[2] PPS [Policy Planning Staff] 28/2, May 27, 1948. Joint Chiefs of Staff Papers, National Archives.

way to cope with this challenge, it was felt, was to assist Asian transformation economically, politically, and culturally so that the countries of Asia would be peacefully integrated into the community of modern nations. The United States could play a unique role in this process because its rapid industrialization, democracy, and Christianity made it the leading example of a country that had successfully organized itself for modernization. Having done so, it could help China, Japan, and other countries of Asia in their transition into modern civilization.

Some of these ideas were already making their appearance in the mid-nineteenth century, when the United States first came into direct contact with China and Japan. This was just the time, it may be recalled, when politics and economics were discussed together as "political economy" and when influential writers such as John Stuart Mill and Friedrich List were raising what probably was the key question of the century: what would be the most appropriate ways to organize human communities for economic transformation? Nationalism provided an answer to some, colonialism to others. In the case of the United States, the question of national political organization was settled by the Civil War, ahead, it would seem, of many other industrializing countries. Thus in the twentieth century, America was in a position to help other countries modernize themselves and to create an international order best suited to the development of the entire world. This was the meaning of Wilsonian internationalism which aimed at creating a cooperative framework not only among industrial powers but also between these powers and the undeveloped regions of the world. Peace was to be based on such cooperation which would, it was assumed, make nations more interdependent and less selfish. The Wilsonian scheme saw its fruition in the Washington Conference (1921–1922). It postulated a framework for American-Asian relations in terms of growing economic interdependence, political stability, and peace. It was assumed that both China and Japan would make progress economically, politically, and intellectually so that in time they would come to approximate the American example, espousing liberal doctrine and contributing to the peace and welfare of mankind. It should be noted that both Chinese and Japanese leaders at the time generally supported this program; they too were searching for the best means for achieving industrial development and political stability. There was no thought at this time that there would be a fundamental conflict between United States-Japanese relations and United States-Chinese relations.

2. *Breakdown in the 1930's, Revival in the 1940's*

The 1930's were a decade when China, Japan, and the United States, instead of acting in the framework of cooperation, clashed. This led to wars between China and Japan, and between Japan and the United States. But these conflicts were more over the means than over the ends of their respective national policies. All three continued to be committed to the same goals as earlier, but they now operated in a drastically altered environment because of the world economic crisis. There was a tendency away from internationalism; all countries pursued nationalistic economic policies with little regard for the stability of the world monetary order, exchange rates, or trade transactions. Japan was not the only offender in this regard, but it was the first major power to destroy faith in internationalism through its aggressive acts in Manchuria and China.

According to the Japanese perception, the familiar world of international economic interdependence, political stability, and mutual good will had been undermined beyond recognition by the economic crisis of the capitalist nations after 1929. If Japan were nevertheless to continue to seek economic development, it appeared that the nation would have to establish an alternative organizing principle. That principle the Japanese believed they found in pan-Asian regionalism. According to this conception, an economically integrated East Asian bloc would eject Western influence and privileges, unite under the leadership of Japan in the name of pan-Asianism, and work together for their collective good.

The Chinese, on the other hand, failed to see anything in such a scheme except Japan's crudely disguised selfishness.

They were no less determined than the Japanese not to let the economic crisis among the capitalist countries affect their own programs for modernization, but they believed they had no choice but to continue to rely on Western countries, especially the United States and Britain, for capital and technology. Sometimes the Chinese were bitterly disappointed, as when America's Silver Purchase Act (1934) drained the metal out of China and caused severe currency fluctuations, or when Britain's proposal for an international program for assisting Chinese industrialization met with lukewarm American responses. Despite all such disappointments, the Chinese were convinced that for them to accept the Japanese idea of an Asian regional scheme would amount to little less than national suicide.

The United States, too, rejected the Japanese scheme, not because it was seen as a formidable threat to American interests in Asia, but primarily because American officials never believed it amounted to very much. Japan, after all, was heavily dependent on the United States throughout most of the 1930's for petroleum, iron, aircraft, automobiles, and other manufactured items. While American officials recognized the Japanese proclivity toward an exclusive economic autarky, they did not think the two countries were after irreconcilable objectives. On the contrary, they felt Japan and America still wanted the same things in order to increase trade, manufacturing, and national product. The United States would, therefore, use varying degrees of pressure to induce Japan to alter its course and desist from a futile attempt to rid East Asia of Western influence. Perhaps the best expression of the American vision was the Atlantic Charter of August, 1941, a declaration jointly made by President Franklin D. Roosevelt and Prime Minister Winston Churchill. In it they called on all nations of the world to effect "the fullest collaboration . . . in the economic field" so that all peoples would "live out their lives in freedom from fear and want." In contrast to Japan's regionalist doctrine, the Atlantic Charter held out the prospect that, in an interdependent world which the democracies were trying to reestablish, all countries would have access "to the trade and to the raw materials of the world" and that everyone would have the right "to traverse the high seas and oceans without hindrance."

This was the definition of an ideal world with which the Japanese could not have agreed more. Even as late as October and November, 1941, some influential officials in Tokyo counseled reconciliation with America on the basis of mutuality and reciprocity as exemplified by the Atlantic Charter. The trouble, it now seems evident thirty-five years after Pearl Harbor, was not that the military and die-hard chauvinists in Japan were completely convinced of fundamental contradictions between the two countries. It was rather that they could not accept the state of uncertainty and drift that had characterized United States-Japanese relations for several years. The two countries had stopped cooperating, but they were not irrevocably committed to a complete rupture in their relations. Such a situation seemed likely to continue even if Japan were to seek ways for reconciliation with the United States, as the tortuous but fruitless negotiations in Washington throughout 1941 indicated. War, for the Japanese, was something that would put an end to uncertainty so that they would *then* seriously begin the task of establishing a pan-Asian order of co-prosperity.

The relevance of these developments to United States-Chinese relations lies in the fact that the United States-Japanese war was not a case of fundamental conflict between two irreconcilable ideologies or diametrically opposed definitions of mutual relations. The war was essentially fought over means, not ends. For this reason China became a symbol of how the two combatants might make use of a third country in solving their problems. For Japan, China was a means to be utilized through coercion and persuasion so that it would contribute to the creation of an East Asian empire of economic development, co-prosperity, and stability, free from Western influence and intervention. The United States, of course, was determined to deny China's use to Japan. In order to prevent Japanese hegemony in Asia and the Pacific, it was imperative to encourage countries and peoples of these regions to look to the vision of a free and prosperous Asia that was to emerge after the war. China would play an important role in such an Asia. The much maligned idea that China was going to be one of the four great powers after the war—maligned because President Roosevelt, despite his rhetoric, did little to encourage the growth of China as a great power—had military and strategic implications. But at bottom it was a perception of China's role as a symbol of Asian development, a development which the United States would promote as an alternative to the Japanese program for regional particularism.

It is interesting to note that wartime American perceptions of China revealed a theme that had also been present in American perceptions of Japan. This was concern lest a victorious China might become as much a menace to peace and orderly development of Asia as Japan had been. As Senator Elbert Thomas said at a State Department meeting in March, 1943, "the more closely we can pin China to a relationship where she will seek guidance of the elder brother which we have become, the better the world will be off and the quicker the resources of China will be developed for the good of the whole earth." Under Secretary of State Sumner Welles, who was at the meeting, agreed, saying that the major question in Asia facing the United States was "what this country can do to make China a stabilizing, peaceful factor rather than a force of future danger."[3] In a similar vein, Stanley K. Hornbeck, political adviser to the secretary of state, told President Roosevelt in November, 1944, "I feel strongly that we must have in mind all the time the need to pursue courses which will keep China within the orbit of our influence instead of driving her out or letting her move away from our influence." This was necessary, Hornbeck said, to prevent a dangerous combination of Asian countries against the West. Roosevelt fully agreed, remarking that "we must be on guard against the possibility of a combination among the orientals in antagonism or opposition to us."[4]

The United States would try to keep China within "the orbit of our influence," just as the Japanese had tried to establish their influence over China. The difference was that the American "influence" was defined in more universalistic

[3] P [Political Committee] Minutes 47, May 12, 1943. Harley Notter Papers, National Archives.

[4] Memo of Roosevelt-Hornbeck conversation, Nov. 24, 1944. Stanley K. Hornbeck Papers, Stanford University.

fashion within the framework of the projected peace aims. According to the State Department's policy committee paper drafted in March, 1944, "The principal American peace aims are to ensure peace and security in the future and to bring about betterment of economic and social conditions." These objectives were to be attained by, among other things, "adoption of international standards and arrangements for the development of dependent areas in the interests both of the dependent peoples themselves and of the world as a whole," and "collaboration between nations directed toward general economic advancement." The strongly economic content of these statements was no accident. It was believed that only through economic development could nations of the world become more interdependent, stable, and peace-loving. Such a development would militate against political particularism and prevent a combination of hostile forces to threaten American security and interests.[5] As the State Department explicitly noted, "American policy is based upon the premise that the economic well-being of a country is a prime factor in its internal stability and in its peaceful relations with other states."[6] Such an assumption would dictate a policy toward China looking toward its emergence as an economically developed and prosperous nation, a member of an integrated and interdependent world, and oriented toward the United States. Unfortunately, American policy toward China during the war became confused because of disagreement among officials over Chinese domestic politics. While all were agreed with the idea of promoting China's modernization, there was growing dispute over which forces in China were "modern" and worthy of American support. The often acrimonious debate between those who insisted on continued support of Chiang Kai-shek and those who urged recognition of Mao Tse-tung as more representative of the Chinese people masked the fact that both sides shared a perception of United States-Chinese relations derived from America's wartime internationalist assumptions.

It is ironic that, while officials disagreed over China, there was remarkable consensus concerning the treatment of de-

[5] PC [Policy Committee] 4, March 29, 1944, Notter Papers.

[6] PWC [Postwar Committee] 52, March 6, 1944, *ibid.*

feated Japan. Here again, the essential ingredient was the idea of development. As George Blakeslee, a scholar who did more than anyone else to formulate basic guidelines for postwar Japan, noted in a memorandum for the State Department, the allies' ultimate objective was "to restore Japan to full and equal membership in the family of nations." More specifically, Japan should be prevented "from again becoming a menace to international peace," but at the same time, "[economic] and financial conditions in the postwar period . . . should ultimately permit Japan . . . to share in the development of a world economy on a non-discriminatory basis, looking toward a progressively higher standard of living." In order to qualify for such treatment, the Japanese would have to establish a government "which will fulfill Japan's international obligations and respect the rights of other states."[7] It so happened that it was much easier for the Japanese than for the Chinese to develop such a government after the war, so that in time the United States came to have greater confidence in the government in Tokyo than in Nanking. That was why, when Kennan returned to Washington after seeing MacArthur, he was able to report that there were good prospects for political stability in Japan, and the National Security Council endorsed the policy of supporting a "middle of the road regime in Japan retaining the spirit of the reform program."[8] Japan, rather than China, was fast becoming America's partner in Asia.

What the above discussion indicates is that, despite the appearance of discontinuities and discrepancies in America's Asian policy, despite the impression of missed opportunities, misunderstandings, and misconceptions in America's dealings with China in the 1940's, there was an underlying pattern that transcended specific circumstances and provided a broad framework for American policy. As indicated earlier, the pattern was derived from the basic question of the nineteenth century, how best to organize nations and the world for modern development. It fell to the United States to take the lead in solving this question in the twentieth century because it became the most advanced industrial power, and

[7] T [Territorial Committee] 357, Sept. 29, 1943, *ibid.*

[8] NSC [National Security Council] 48/1, Dec. 23, 1949, Joint Chiefs of Staff Papers.

because it had always been deeply involved in underdeveloped regions of the world even while shunning entanglement in Europe. America had especially been interested in attempts to organize Asia for peace and prosperity. These attempts provided the basis for guiding United States policy during the Second World War and thus defined America's approach to postwar Asia. It was immaterial, in such a perspective, whether the United States was dealing with Japan or China.

3. *The Cold War as an Interruption*

When the Second World War ended, the same objectives of United States policy could have been pursued both in China and Japan. Unfortunately, the cold war intervened in such a way that bilateral United States-Japanese ties, rather than trilateral United States-Chinese-Japanese relations, developed. There are those, it is true, who argue that the cold war was itself a logical culmination of America's efforts to modernize the world in its own image. Some go even farther and say the cold war in Asia was but an extension of American imperialism that had sought economic domination over the world. Professor Fairbank implies some such connection when he says, "Imperialist aggression in China inspired the rise of nationalism, and revolution has been the result." I would argue that nationalism more often antedates than postdates imperialism. Be that as it may, in the context of this paper it seems possible to say that the postwar strategy of cold war confrontation with Soviet communism served to deflect American policy in Asia away from concern with economic development and political change to preoccupation with reducing Soviet strength and influence in the region. America's resources would be devoted to containing Soviet expansionism, real and imagined, and the degree of American economic and technical assistance of underdeveloped countries would be determined by the overall dictates of an anti-Soviet strategy. For this reason, the bulk of economic assistance programs during the first decade after the war went to Europe and the Middle East. East Asia was usually put at the bottom of priority lists because it was felt that other regions of the world were more likely to be theaters of American-Russian confrontation. When the National Security Council formulated a new Asian policy in December, 1949, in view of the Communist takeover of China, economic development *per se* was given little emphasis. The United States, it was decided, should "carefully avoid assuming responsibility for the economic welfare and development of [the Asian] continent."[9] Such language was a far cry from traditional sentiments and objectives. These were now subordinated to more political and ideological goals of resisting communist expansionism. In such a situation, it was not surprising that the United States should have placed stringent restrictions on economic transactions with the People's Republic of China, or that it should have considered policy toward the new regime in Peking primarily in the context of combating the Soviet Union. Only in Japan was the United States able to continue the policy of encouraging economic development and political reform.

With the passing of the cold war confrontation with China in the early 1970's, the stage was set for picking up the thread in United States-Chinese relations that had been lost in the complications of the cold war for over twenty years. The two countries were ready to resume economic transactions which, from the American point of view, were to be a fulfillment of century-old hopes and expectations. Even more important, United States-Chinese rapprochement would not be effected at the expense of United States-Japanese ties. This was completely in line with the traditional approach. It would seem that in the 1970's the United States has finally returned to the old objective of promoting Asian development and stability on the basis of peaceful relations both with China and with Japan. If economic and cultural ties between America and China were to grow deeper in the years to come, one might well see in such developments the fulfillment of one of the most durable ideals for which this country has stood.

AKIRA IRIYE
Professor of History,
University of Chicago

[9] NSC [National Security Council] 48/2, Dec. 30, 1949, *ibid.*

Economics and Technology in United States-China Relations

INDUSTRIALIZATION IS A world-wide movement that spread first from England to Europe, America, and Japan. Since World War II, the rate of diffusion of the industrial system has accelerated particularly in Asia, Latin America, and parts of the Middle East. Perhaps most significantly, the People's Republic of China, one-fifth of mankind, has by now experienced a quarter-century of sustained economic development. The level of output of Chinese modern industry today is only about a decade behind that of Japan and is increasing at ten per cent or more a year, a rate that will double its size every seven years.

Popular mythology has it that China has achieved this performance on its own without significant assistance from outside the country. Where other recently successful industrializers such as South Korea, Brazil, and Iran have relied extensively on external economic and technical assistance, China, it is said, has gone it alone both by necessity and by choice.

If this portrait of Chinese independence were completely accurate, the subject of this essay would be of only limited interest. American economic relations with China, after all, are only a part of China's economic relations with the world as a whole. If China's economic relations with the world have not had much of a role in Chinese economic growth, economic relations with America both present and future would be only a part of something which in its entirety was not very significant.

Part of the confusion over whether China has or has not pursued an "independent" economic development program arises because writers seldom define what they mean by "independent," "autarkic," or "self-reliant," the last being the term preferred by the Chinese themselves. To some these terms refer to a lack of Chinese dependence on foreign aid and credits in their investment program. To others it is the Chinese desire to limit the need for imported commodities that is essential. Many also suggest that China has created its own unique development program rather than copy the nineteenth-century experience of the West. There are even a few people who take the extreme position that China is developing most of its own technology. Let us take up each of these definitions of independence in turn and explore their implications for United States-China relations.

1. *The Degree and Nature of China's Self-sufficiency*

It is in the area of foreign aid, foreign investment, and other external sources of financing that the case for China's claim to self-reliance is most compelling. China did receive foreign aid from the Soviet Union in the first half of the 1950's; the amount allocated to economic projects totaled roughly one billion dollars U.S. But there was no Soviet financial aid of significance after 1955, and China had repaid all Soviet credits of any kind by 1965. In 1974 China ran a large balance of payments deficit of around one billion dollars and this deficit persisted into 1975 although at a level about half that of 1974. Deficits in both years were financed by commercial loans, however, not by anything that could reasonably be called aid. Because the United States and China have yet to settle the issue of Chinese frozen assets in the United States and United States private claims against China, and because the Chinese discriminate against United States banks in the absence of normalized diplomatic relations, American banks were not directly involved in the financing of these deficits.

There has been no new private foreign investment in China since 1949 and what existed at that time has all been confiscated. There is no prospect for any shift in this hostility toward foreign ownership of Chinese property anytime in the foreseeable future. The source of this hostility is political and arises both out of Marxist-Leninist ideology and out of China's unhappy historical experience with foreign investment in the nineteenth and early twentieth centuries. Other countries' leaders, however, have felt a deep hostility toward foreign investment and yet their nations still permit such investment and often indeed encourage it.

China has treated foreign investors cavalierly in part at least because there was no compelling reason to do otherwise. Executives of some of the large multinational companies occasionally talk as if nations simply cannot afford to finance the huge expenditures involved in oil exploration, for example, and hence these nations are forced to turn to the multinationals. For very small countries this may be the case, but for China the contention is absurd. In recent years China has been investing in her own economy annually over $50 billion U.S., an amount that is over six times greater than the total American direct investment per year in the entire world. Even if the United States were to channel all of its direct external investment toward China, therefore, the amount would meet only a fraction of China's current investment needs.

If one turns from investment to foreign exchange, China's resources are more modest but still far greater than any amount available from foreign private sources under the most favorable assumptions possible. Annual American direct investments in all of Asia including Japan amount to a few hundred million dollars. Chinese foreign exchange earnings per year from exports, in contrast, are running at a rate of about $7 billion U.S. In short, China's current demand for both investment funds and foreign exchange can only be met from China's own resources. Funds available from abroad, whether from the United States or elsewhere, are such a tiny fraction of what China needs that there is little temptation for Chinese leaders to repress their opposition on political grounds in order to tap these foreign resources.

The question of China's dependence on foreign trade in general and trade with the United States in particular is more complex than the issue of dependence on foreign financing. The Chinese themselves dispute some of the more popular notions concerning the meaning of self-reliance.

> Some foreign friends have raised the question that since China carries out the policy of self-reliance, following the development of China's national economy, her foreign trade would diminish in importance. We dispute this. . . . When China will have accomplished the comprehensive modernization of her agriculture, industry, national defense and science and technology, she will still need to import new products and techniques from other countries. Following the development of production, we shall be able to offer more and better products for export. Therefore there is no doubt that our foreign trade will continue to grow.[1]

The above is certainly not the statement of a planner dedicated to gradually cutting back all economic ties with the outside world. It could, in fact, be the statement of a trade minister in a nation dedicated to getting the most out of the international division of labor.

Is China's trading pattern with foreign countries substantially different because China has a Communist government dedicated to avoiding dependence on the outside world, a dependence that it is felt could jeopardize the future of the nation? The answer is yes, but to only a limited degree.

The two major determinants of China's foreign trade are China's size and the nation's desire to industrialize. In population China is the largest nation in the world and in land area it is one of the largest. In large nations fewer products (as a percentage of total national product) enter into international trade than in small nations. The United States, for example, imports less than four per cent of all the goods and services it uses in a given year, and the Soviet Union about three per cent. Small countries, in contrast, import on the average around twenty per cent of their needs. The basic reasons for this contrast between large and small nations is

[1] Chung Wen, "Rely Mainly on Our Own Efforts while Making External Assistance Subsidiary," *China's Foreign Trade*, No. 4 (1975): p.5.

straightforward. A nation with five million people cannot build an industrial sector capable of producing all needed products at home. In many industries, an efficient scale of operation would produce far more than could be used by the small country itself. Thus the logical step for such a nation is to specialize in producing some items at an economical scale, exporting the surplus, and importing other requirements. In a nation of 900 million, in contrast, the market is large for all but certain specialized items even when per capita income is low. Furthermore, the costs of transporting a product from abroad are likely to be higher in a large nation. The cost of shipping coal or cement from China's coast into the interior, for example, could be prohibitive. It makes much more sense to build cement plants and develop coal mines (if possible) in the interior. Finally, China like most large nations is rich in natural resources. For only a few natural resources does China have to resort to imports because the commodity is completely unavailable within its own territory.

China today imports about five per cent of its requirements each year. The figure rises slightly in the midst of a development drive and falls in a major recession such as that of 1960–1961. Clearly this low percentage mainly reflects China's size, but a case can also be made that Chinese government policies have kept the percentage of imports lower than it might otherwise have been. People who trade with China frequently remark on Chinese unwillingness to design and market their products in a way that would make them appealing to foreign buyers. Inappropriate brand names (e.g., "White Elephant batteries") are only a small part of the problem. More serious is the inability of the typical foreign buyer to know from year to year whether the Chinese will have any of a particular item to sell. If he gears up an advertising program based on one shipment, he may not be able to get another shipment for a long time and the advertising expenditure will have been wasted. There are many other problems of this kind. Most of all they reflect a planning and management system that is geared to China's domestic market and treats foreign trade and marketing as a function separated from the main task of production. It is not that Chinese officials are trying to hold down trade so much as it is that those with the power to affect the outcome are only occasionally concerned with expanding that trade. Greater effort to develop cash crops, for example, might make possible a much larger export surplus of these commodities, but Chinese agricultural planners and scientific research stations put most of their energy into raising grain output for domestic consumption.

Because China's trade is small (relative to gross national product), it does not follow that trade is unimportant to China's economy. To the contrary, a case can be made that as the trade ratio is brought down below some hypothetical maximum, the trade that remains is reduced to what is absolutely essential. Certainly this has been the case with the People's Republic of China. In the 1950's, for example, China imported hundreds of complete plants from the Soviet Union and Eastern Europe along with thousands of technicians to help set them up and get them into operation. Without these imports, the first five-year plan (1953–1957) and its industrial development program centered on steel and machinery would never have got off the ground. And without the development of steel and machinery in the 1950's, China would have been unable to carry out the large investment-industrialization programs of the mid-1960's and early 1970's.

By the first half of the 1970's China's industrialization efforts no longer depended on imported plants and equipment to the degree that they did in the 1950's, but the dependence in key areas was still substantial. Perhaps the least essential of China's imports in the 1970's was the grain purchased from Canada, Australia, and the United States. The amount purchased, five to six million tons a year, was a tiny fraction of total Chinese grain supplies of 250 to 280 million tons annually. The cost of eliminating grain imports would mainly have been felt by Chinese farmers who would have had to increase their deliveries to the state. Increased deliveries to the state would have hurt the farmer's incentive to raise production and would have been an added problem for an already overburdened transport system. Thus the elimination of grain imports would not have led to starvation or some other drastic result, but the price of elimination would not have been cheap.

Certain other imports are essential because China is unable to produce significant quantities of these commodities at home. Rubber and long staple cotton are two examples in this category. Even more important are the imports of complete plants for the chemical and steel sectors. China is capable of putting up, on its own, small fertilizer plants producing ammonium bicarbonate, but large efficient plants manufacturing urea, a superior form of fertilizer, are beyond China's present technical capacities. Similarly, with certain kinds of high quality steels the choice facing China is to import either the finished product or the plants to produce that product. As time passes, of course, China will eventually learn how to produce its own urea plants and advanced steel rolling mills, but by then Chinese industry will be on a new frontier requiring imported equipment of an even more advanced type. Some of the new technologies required for this equipment may be developed within China, but China will remain heavily dependent on imports for the most advanced technology. Most of China's small science and technology establishment will have to devote its energies either to adapting imported techniques to Chinese conditions or developing essential items not available by purchase from abroad. This latter category includes everything from nuclear weapons to plant varieties suitable to China's soil and climate. Chinese scientists and engineers, therefore, will have their hands full without taking time out to reinvent the wheel (or Boeing 707's).

2. *The Prospects for United States-China Trade*

Where does trade with the United States fit into this picture? To begin with, the United States is not the world's only source of advanced technology, but it is an important source. Some kinds of equipment are readily available, often at lower prices, in Japan and Western Europe. Other kinds, however, are not so readily available elsewhere. High quality computers and commercial aircraft are two of the more obvious examples. Nor is the United States solely a source of advanced technology. The number of nations with a surplus of agricultural products is dwindling and North America and Australia with their rich supply of arable land are among the few reliable sources of food exports remaining.

The first year of significant United States-China trade since the imposition of the embargo in 1950 was 1972 (see table 1). The composition of United States-China trade since then has largely reflected the two main American advantages just described. Two of the biggest single sales by the United States to China were ten Boeing 707's and eight ammonia-urea plants by the M. W. Kellogg Company. After China's poor harvest in 1972, the United States also became a major exporter of agricultural products to the People's Republic of China, pushing total United States exports to nearly one billion dollars U.S. and making the United States in 1973 and 1974 China's second largest trading partner. The euphoria in some American circles surrounding this rapid rise in trade with China was brought down to earth in 1975 when the Chinese cut back sharply on purchases of farm products, and the United States traders bore the brunt of the cut.

The position as of 1976 is that China discriminates against American products whenever the cost of discrimination to China is not great. The purpose is to put pressure on the United States to normalize diplomatic (and hence trading) relations with the People's Republic. When normalization has been achieved, American exporters will be able to compete on more even terms with Japanese and Europeans, and American exports to China will rise, but not as dramatically as some people expect. In the meantime, the United States continues to sell considerable quantities of machinery and equipment even in the absence of normalization.

Clearly $300 million U.S. worth of exports does not make the difference between success and failure in China's eco-

TABLE 1
SINO-AMERICAN TRADE (1971–1975)
(unit: million U.S. dollars)

	Exports to China	Imports from China
1971	–	4.9
1972	63.5	32.4
1973	740.2	64.9
1974	819.1	114.7
1975	303.6	158.3

Source: *U.S.-China Business Review* **3,** 2 (March-April, 1976): p. 30.

nomic development plan. The total is small relative to a total investment level of over $50 billion U.S., and many of the items are available elsewhere even if on less favorable terms. Thus in no sense does China depend vitally on economic relations with the United States. The United States, of course, is even less dependent on trade with China. Both countries, after all, got along without trading with each other for two decades prior to 1972. On the other hand, Chinese purchases of the products of Western Europe, Japan, and the United States taken together are crucial to key parts of the nation's development program. If one of these sources were not available for whatever reason, China would be that much more dependent on those that remained. In that sense the continuation of United States-China trade is of considerable importance to the People's Republic.

Finally, trade is not the only way for China to acquire the latest achievements of modern science and technology. If a nation has trained scientists and engineers, these can be used to bring in new techniques in a variety of ways. Scientists trained abroad often bring knowledge of advanced technology with them when they return home. China's nuclear and missile programs, for example, depended heavily on scientists trained at the Massachusetts Institute of Technology and the California Institute of Technology who were hounded out of the United States in the McCarthy period, together with a later group of nuclear scientists trained in the Soviet Union. Other scientists subscribe to the best academic journals and keep up in that way. Still others participate in exchanges with the United States and other advanced nations, exchanges that normally involve visits to some of the best laboratories and industrial plants in those countries.

From an American point of view, perhaps the central issues in this transfer of technology from the United States, Europe, and Japan to China is whether this transfer will put pressure on China to open up the country to more outside influence and whether access to American technology is important enough to encourage China to make a major effort to maintain good relations with the United States.

We are no longer so naive as to believe that scientific progress is possible only in a free atmosphere where scientists can communicate with each other and with their colleagues in other countries. The experience of the Soviet Union has taught us otherwise. Nor can one make much of a case for the proposition that trade by itself necessitates large-scale and in-depth foreign contact. Foreign traders traveling to China to do business seldom even see the factories for which their equipment was purchased. When complete plants are set up, there is a period, often of several years, when foreign technicians must reside at the factory site. But as soon as these plants are finished and in operation, the technicians are packed up and sent home. The Chinese (like the Russians) do not even allow the technicians to stay during the first few years of operations, although in most cases this step costs the Chinese considerable amounts of lost output and inefficient production. Ridding the nation of an alien element is deemed more important than somewhat greater economic efficiency.

Thus neither trade nor technology transfers are by themselves likely to push China into a more open and dependent relationship with the United States or other parts of the advanced industrial world. China's minimum requirements can be met with contacts much more limited in nature. But even though more intimate contact with the outside world is not necessary for economic and technical progress, still China's scientific and economic progress will almost certainly be more rapid with such contact than without it.

Do the Chinese leaders themselves share the view that there is some tradeoff between economic-scientific progress and the degree of openness of the society to the outside world? Certainly there is no unanimity of opinion on the subject. One major issue of the debates between "right" and "left" during the Cultural Revolution and most recently during and after the fall of Teng Hsiao-p'ing has been precisely this subject. The "right" has regularly been accused of slavish imitation of foreign methods, excessive admiration of American technology, and the like. Although such debates are always marked with a good deal of hyperbole, there is little doubt that real differences of opinion exist over this issue.

It is not that any of China's leaders want to open up the nation to foreign influences in any dramatic way. All would probably consider this a move fraught with both political and economic risks. Political unity and control in China have been possible, in part at least, because ideas competing with those promulgated by the Communist party have been systematically excluded. And a high rate of savings and investment has been easier to sustain than if the Chinese people were constantly being bombarded by images of the consumer culture of Japan and the West.

But there is a real difference of opinion over whether China has a lot or a little to learn from foreign technique. To the group that has been labeled everything from "moderates" or "pragmatists" to "capitalist roaders" depending on the point of view of the writer, a group that includes most of those responsible for managing the economy, or what is now referred to as the "gang of four," the advantages of greater contact with technological developments abroad are apparent. To the "radicals," "left," which includes many former leaders in the cultural and educational spheres together, paradoxically, with the former political heads of China's most advanced city, foreign techniques are seen as undermining China's will and ability to develop its own unique path to a modern society and economy.

To Americans it is almost an article of faith that we would be better off with the former group than with the "left" in charge of the world's largest nation. But if the moderates combined a desire for more advanced technology with a decision to seek rapprochement with the Soviet Union, the advantages to us of one group over the other would be far less apparent. Perhaps, given our ignorance of the full consequences of the ascendancy of one group over the other in China, it is just as well that we have little ability to influence the outcome. All Americans can really do in this sphere is to retain a willingness to share the fruits of our economy and technology with those of China on terms advantageous to us both, and to hope that these exchanges will contribute in some small way to improving relations between our two countries in all spheres of activity.

DWIGHT H. PERKINS
Professor of Economics,
Harvard University

Scholarly Exchange with the People's Republic of China — Recent Experience

In 1972 EXCHANGES of Chinese and American scholars in all areas of the sciences—physical, biological, engineering, as well as the social sciences and humanities—were re-established. Since then some one thousand American scholars have been to China and about three hundred Chinese scholars have visited the United States. The Americans who have gone to China have been primarily from the biomedical and physical sciences, with a smaller number in the social sciences and humanities. Efforts to provide a balanced program representing American interests have not been particularly successful, as the People's Republic of China is highly selective as to whom they admit. For example, it has been difficult to arrange programs on contemporary Chinese society. Admitting social scientists and China specialists is probably not considered to be useful and, in fact, may be perceived as potentially dangerous by the Chinese.

The fields represented by Chinese scientists traveling to the United States reflect China's interest in exchange programs, and that is to review Western advanced science and technology. Visiting delegations have been in such highly applied fields as computer technology, high energy physics, telecommunications, solid state physics, laser technology, petrochemicals, and industrial automation. In addition, China has sent targeted basic research groups. For example, in the biomedical field visiting parties have been concerned with such areas as tumor immunology, molecular biology, and pharmacology, and in agriculture there have been groups in plant breeding and photosynthesis. These groups have included middle-aged and younger scientists, with a few older scientists trained abroad in preliberation times. Most visitors have never been out of their country and must find the experience revealing, if not jolting.

The Committee on Scholarly Communication with the People's Republic of China was already six years old when the exchange program began in 1972. The CSCPRC had been formed in 1966 by some far-sighted individuals, well ahead of their own government, who foresaw future developments and wisely organized a committee to promote and operate exchanges with China. The CSCPRC is a committee of several learned societies—the Social Science Research Council, the American Council of Learned Societies, the National Academy of Sciences, and the National Academy of Engineering. This broad sponsorship supports the Committee's claim of representing a goodly section of American scholars. The CSCPRC sponsors about twenty-five per cent of the scholars who travel to China. Perhaps China is wise in not giving the CSCPRC a monopoly in sending American scholars to China. The CSCPRC attempts to reflect the priorities of the American scholarly communities. But by dealing with the CSCPRC for some programs and inviting scholars independently as well, China can maintain a flexible operation to pursue its own priorities. It is interesting, however, that China does use the CSCPRC as a host organization for over ninety per cent of the Chinese scholarly groups visiting the United States. This may be because the CSCPRC, through its sponsoring organizations, has good access to universities and research laboratories.

The current exchange format is a survey visit of three to four weeks. Americans going to China can expect to visit the major laboratories or institutes in their fields, as well as museums, communes, and general tourist attractions. By and large, American visitors have no idea what they are going to see in China before they arrive in Peking. They may have requested itineraries in advance, but it is rare to receive such information beforehand. In contrast, Chinese visitors have such a thorough knowledge of work in the United States in their fields that they are prepared to make very specific requests as to whom they wish to see and where they wish to go. For the most part, the CSCPRC responds to their requests. Furthermore, they have an intimate knowledge of United States publications and have read extensively before they arrive. They are prepared to make the most of their visits to the United States.

American scientists visiting China are given access to scientific data in varying degree and are shown scientific equipment. Discussions with Chinese scientists at the institutes and even at one's hotel are frequent. Chinese scientists prepare lectures for their visitors, but since this kind of communication is unusual for Americans, they are frequently taken aback by the political content of presentations which include comments on the political and ideological aspects of China's scientific and technological work. Americans are disappointed at the lack of follow-up communication after they return to the United States. It is difficult to establish any kind of continuing relationship with the people one has met, or even to send and receive reprints.

Scholarly exchanges, therefore, are superficial at this stage. Visits provide a sense of the level of effort, the quality of work, and the nature of training. But there are no opportunities to engage in serious communication or to cooperate in joint projects. Visits whet the appetite but do not satisfy. This must be true for Chinese visitors to the United States as well. Survey visits here cannot provide details of how scientific work is conducted. The future evolution of the program will have to include longer term exchanges if either country is to obtain real benefits.

There is no question that, from the points of view of the respective governments, the immediate basis of exchange is political. Exchange programs are considered in the overall context of political relations and of global power politics.

They are a symbol of the rapprochement between our two countries. In a sense, then, the exchange program is a means of sending a message to each other and to the rest of the world, particularly the Soviet Union, that the world's most populous country and the world's most technologically advanced country can cooperate. This cooperation can either grow or not. So symbolism is important and may be more important to our government than content.

There are, of course, other motivations on both sides. Chinese scientists want to see, at first hand, the work they have been reading about in our scientific literature. China also would like to show her advances in particular areas of science—earthquake prediction, for example. Americans visiting China want to review Chinese work in all fields. We have been cut off from a generation of Chinese scholars: Who are they? How good are they and how are they trained? What role do they play in their own country? How does the mixture of science and politics in China work out in practice? We have heard that graduate education was abolished during the Cultural Revolution. Have they been able really to train physicists and mathematicians using these radical education plans?

China's scientific potential is immense. In time, China will contribute to the scientific and technological solutions of global problems: food production, energy, disease, natural disasters, and Third World development. There are also "big science" projects which will begin in the next few decades and which no country by itself can afford. China's participation could be important to these world efforts. It may be in our interest to help China "take off" now, by a one-way transfer of technology, so that in the long term she may contribute to efforts of world dimension.

Americans also are motivated by simple China euphoria—the great desire to see the country and the people. Thousands of Americans want to know how they can go about getting there.

After four years of exchanges, those who have supported the programs are beginning to try to evaluate what has happened. Despite the modest scale, the initial stage of exchange programs could be considered successful. We have seen a cross section of Chinese scholarship, especially in science and technology. We have learned something about Chinese education and training and have some idea of China's accomplishments. More important, we have established a working relationship with China's scientific organizations. Although relatively few Americans have been to China, what they have learned has been widely disseminated through trip reports published by the National Academy of Sciences, professional journals, university presses, and the like. The Committee on Scholarly Communication with the People's Republic of China publishes yearly five reports by its delegations to China.

The exchange program has shown a slow but important evolution. Chinese visitors to the United States now give lectures on their work and on the state of their field in China. When they first began coming to the United States years ago, they sat quietly in our laboratories and received information. Visitors to both countries now are staying longer periods in fewer places, thus providing better opportunities for discussion than did the earlier, more hurried visits. In some fields we have had repeat visits—the beginning of continuity in relationships. So, in some ways, there has been progress.

On the other hand, Americans tend to be impatient and many feel that the first get-acquainted stage has lasted long enough. We would like the program to evolve more rapidly from scientific tourism to significant interchange. China's representatives here in Washington, as well as the people Americans see in Peking, tell us that any significant developments are tied to the establishment of diplomatic relations. There are other problems, of course. The greatest one may be the difference between our two societies and how that affects the conduct of scientific work. Americans have competitive ways and independence which is a completely different style from that of Chinese scientists who work not for themselves, but within a group. Cooperation on projects may be difficult.

There is an imbalance in the exchange relationship and that is a problem that must be addressed. Many laboratories in universities and industrial firms have become impatient with repeated visits by Chinese delegations with little progress in the form of business relationships, or of their own people going to China, or of learning about Chinese work in their own fields. The tough question, "What are we getting out of it?" is being asked, not so much in academic circles, but elsewhere in industry and in the Congress. The questions now being asked about the exchange program with the Soviet Union after twenty-five years of experience are being asked earlier with regard to China.

The problem of Chinese discrimination against the social sciences and the humanities in exchange programs is a serious one. Perhaps China views social scientists and China specialists as potential agents. They know a great deal about the country and, therefore, could learn a great deal quickly. From such sources, valid and educated criticism would be widely broadcast. China perceives little benefit in receiving such visitors who probe their society and their system, whereas they derive enormous benefit from visitors in technical areas. There is no question but that this dichotomy in Chinese interest creates tension in our own scholarly community.

The future of scholarly exchanges is uncertain. The new leadership in China seems interested in industrial development and Western technology. This mean an increase in exchanges, as well as trade with the West. Should diplomatic relations be established in the next few years, this too may provide a better climate for exchange development. We expect, for example, that a language exchange program would then be established. But would we then see longer-term exchanges? Would there be opportunities for joint conferences? Would we have young Chinese scientists here for periods as long as a year? I think so, for pragmatic reasons. In my opinion China's Cultural Revolution set the training of young scientists back a generation. If China wanted to use American resources in the most effective way possible, she would send young scientists to our laboratories for training. For this reason we may see exchanges evolving beyond scientific tourism to more substantial programs.

In conclusion, consider that in four years we have gone from a state of no contact to a superficial level of contact by a few people. Perhaps we could not have expected more in the initial stage. It may be wise to be pragmatic in our evaluation of the exchange program, but at the same time take the long-range view of integrating benefits over the next ten or twenty years. If we have a predominantly one-way transfer of technology to China now, the return to us and to the world twenty years from now in the way of Chinese contributions to world problems of food, disease, environmental pollution, and energy may be substantial. At the same time, we must consider how to press for programs in areas of our own interest.

There are 900 million Chinese people with a government interested in scientific and technological development. They certainly will soon operate at world levels in many fields. They need access to our universities and laboratories to proceed efficiently with their development. It is in our interest as Americans to know the new generation of Chinese scientists and to establish friendly relations. These are strong motivations for both sides. For these reasons, I believe exchange programs will develop and the years ahead will be exciting ones.

FRANK PRESS[1]
Chairman, Department of Earth and Planetary Sciences, Massachusetts Institute of Technology

[1] I am indebted to Anne Keatley, Alex DeAngelis and Mary Bullock of the CSCPRC for their help in preparing this report.

Comment

IN PONDERING OVER the above discussion, one is struck by the startling realization that the issue of Sino-American relations received no serious consideration in the presidential campaign of 1976. But that need not lead us to conclude that the matter is unimportant: aspiring politicians do not always talk about the most important things in the world.

This matter has a special timeliness, given the changing of the guard in both Peking and Washington. China has now entered the post-Mao era, and one can discern a new pragmatism dominant in Peking. In Washington, with the shift from a Republican to a Democratic administration, according to campaign promises there will be more openness, and a new morality, in our foreign affairs. The moral factor could introduce complications.

What of the future of Sino-American relations? Professor Press's article sketches for us an unpromising prospect. Reference to history is helpful in this regard: as the Chinese proverb has it, *ch'ien ch'e shih chien*—"the cart ahead is a mirror." And Professor Fairbank offers a view of the cart ahead and the route it has traversed. Noting the Sino-American cultural gap, and an intellectual gap, he underlines the need for understanding by American policy-makers of China's past. Here I add a point: it is of course entirely evident that the failure of understanding has been mutual;

the Chinese have failed in their understanding of us as much as we have failed in our understanding of them.

Mr. Fairbank offers some contrasts. He says that China is the "oldest continuous polity." Yes, but China has gone through a big variety of geographical manifestations over the centuries. He says that in contrast the United States is the most recent polity to achieve Great Power status, and that it is in the flux of a technological civilization "whose problems even defy listing." Let us offer, however, a blanket characterization of the so-called Western civilization that both shows a parallel to China in the first half of this century and relates it to Mao Tse-tung's late thinking: this is a civilization experiencing major institutional tensions—particularly in the economic sector.

Mr. Fairbank draws a parallel between the two, finding both to have been maritime powers; but here too there is a difference: despite the Chinese maritime adventures of Ming times, Ch'ing (Manchu) China felt self-sufficient. The Emperor Ch'ien-lung expressed that feeling in his communication to King George III in 1793:

> The Celestial Empire possesses all things in prolific abundance and lacks no product within its borders. There is therefore no need to import the manufactures of outside barbarians in exchange for our own products.

As the "Middle Kingdom," China deemed itself powerful, rich, self-sufficient, with no need for others. It could indulge in "self-reliance." And at that same time, a handful of Americans possessed a rich continent to exploit, and would before long cast their gaze farther westward, across the Pacific.

Now, in the latter half of the twentieth century, there is a clear contrast with the estimate of Ch'ien-lung: the People's Republic of China is judged to be a have-not nation in terms of economic development; it says as much, and purports to belong to the developing Third World. Here the political factor is to be considered. As stated by Mr. Fairbank, Chinese rulers of today, like those in former times, aspire to rule *all* of China—and this includes Taiwan. But of course the changes in the map of China over the dynasties show that the country's rulers have not always fulfilled their aspirations regarding political dominion. Yes, there is the One-China doctrine. But *what* One China? Chou China? Han, Sung, or Ming China? No, China's rulers usually aspire to have a China of greatest territorial dimensions—a Mongol or a Manchu China. The rulers of today's China would include Taiwan within the country's borders. But the practical question is: how shall Nationalist China and Communist China be reunited?

The matter is complex and difficult to resolve, for a variety of reasons. There is the legal factor: the United States is tied to Taipei by the treaty of December, 1954, and, although there is provision for abrogation of the treaty by one year's advance notice, the United States is concerned for the sanctity of treaties. There is the political factor: the United States' concern for its "credibility," and worship of the doctrine of self-determination—a very important factor, especially if "morality" enters in. There is the military factor, given less publicity but present in the equation: considerations with respect to the national security of Japan, South Korea, and the Philippines. And finally, there is the economic factor: the United States trade with Taiwan is some seven times the size of our trade with the People's Republic.

Professor Fairbank in conclusion points up the distressing situation in which we find ourselves as regards Taiwan. The hard fact is that the United States made the mistake of assuming, with Secretary of State John Foster Dulles, that Communism in China was "a passing phase," and now it is difficult to extricate ourselves from the situation we created by our miscalculation. So we confront that challenging question, "How?"

Professor Iriye makes a suggestion that is pertinent to our subject: add Japan to the bilateral relationship. He points out that Japan and the United States acted as allies against the People's Republic of China during two cold-war decades. Note here that Japan established regular diplomatic relations with the People's Republic in 1972, and has a bigger trade with that country than has the United States. We thus confront a changed situation: in the trilateral relationship, Japan is now closer to the People's Republic than is the United States.

Mr. Iriye, like Mr. Fairbank, adduces the concept of "continuity." Here I would voice a caveat: in this post-World War

II era, there is much that has been *dis*continued—for example, colonialism, the role of imperialism, Western domination. Deep new trends are quite visible. The People's Republic of China and Japan are undergoing change in a changing Asia; the United States confronts change in a changing Occident.

Mr. Iriye notes a nineteenth-century issue worth our focusing on. He says that nineteenth-century writers like John Stuart Mill and Friedrich List sought to discover what would be the most appropriate ways to organize human communities for economic transformation, and that Wilsonian internationalism aimed at creation of a cooperative framework "not only among industrial powers but also between those powers and the undeveloped regions of the world." This reminds us that there is now something in the nature of a confrontation between three world combines—the capitalist bloc (we use the term "the industrialized democracies"), the socialist bloc, and the Third World. The times indeed demand cooperation; but the threat is of greater conflicts ahead.

Referring again to history, note an example of conflict. Mr. Iriye states that the Sino-Japanese and American-Japanese wars of the 1930's and 1940's "were more over the means than the ends of their respective national policies. All three [countries] continued to be committed to the same goals as earlier, but they now operated in a drastically altered environment because of the world economic crisis." I disagree as to the nature of the differences. What were the goals of those countries in that period? China aimed at ridding itself of what it viewed as the vestiges of imperialist infringements, and at resuming the role of Middle Kingdom in terms of "Wealth and Power." Japan, as determined from the time of the Manchurian incident of 1931, aimed at achieving domination in East Asia, with domination to include an economic manifestation—the "Greater East Asia Co-Prosperity Sphere." The United States, in an extension of the 1898 victory over Spain that brought it dominion over the Philippines, had the political goal of maintaining the balance of naval power in the Pacific and the economic aim of maintaining access to the market of China in particular—to Carl

Crow's "400 Million Customers." Be it remarked that Mr. Iriye's own subsequent definitions establish that the two wars were fought over means *and* ends.

Mr. Iriye sees a basic American interest in organizing Asia for "peace and prosperity" diverted after 1947 by cold-war considerations that dictated a "preoccupation with reducing Soviet strength and influence in the region," but then he sees in the early 1970's the resumption of Sino-American economic relations which, "from the American point of view, were to be a fulfillment of century-old hopes and expectations"—but not at the expense of United States-Japanese ties. But note that Chinese, American, and Japanese aims are still in no little conflict.

This all leads very logically to considerations set forth by Professor Perkins. Ch'ien-lung saw China as self-sufficient; the country's present-day rulers see it as economically deficient. Mr. Perkins tackles the problem of the economic development of the People's Republic of China. He holds that "China's current demand for both investment funds and foreign exchange can only be met from China's own resources." "Only"? Exclusively? No foreign aid, complete self-reliance, autarky?

That was of course Mao Tse-tung's position, and it is the present pressing issue in Peking. But Professor Perkins himself notes the aid rendered to the People's Republic by the USSR in the 1950's; and it is not to be denied that the Chinese economy advanced the more rapidly by reason of that aid, as was duly remarked by various Chinese leaders, including Mao Tse-tung, at the time. Note here certain historical parallels. The United States made use of foreign aid, in terms of technology and capital, for its industrialization in the nineteenth century, and Japan had recourse to foreign technology for its modernization process in the Meiji era. The USSR utilized much foreign aid during the period of its first Five-Year Plan, and does so again at the present time. Where the USSR is currently going into debt to the tune of tens of billions of dollars in order to advance its economic development the more rapidly, shall the People's Republic choose to rely solely upon its own efforts, to indulge its pride in autarky? Is this the advance in Chinese thinking from the time of Ch'ien-lung down to Mao Tse-tung?

Note here the aim set forth in Chou En-lai's speech to the Fourth National People's Congress in January, 1975. The People's Republic of China shall be transformed, he said, into "a powerful, modern socialist state by the turn of the century." It is requisite that we look to that debate between "radicals" and "pragmatists" discussed by Mr. Perkins. With the death of Mao Tse-tung on September 9, 1976, the issue was joined, and already it is apparent that the pragmatists have won out over the radicals at this stage. The stress is now on Party unity—discipline; on stability—political continuity; on production—economic progress. I am prepared to make a prediction: the People's Republic of China will probably follow in the footsteps of the United States and Japan in seeking foreign technology and industrial equipment and, yes, foreign capital if available, to hasten its development in a world that is in a growing crisis; and this even though it would thus also be seen retreading the footsteps of the "revisionist" Soviet Union.

Professor Press considers scholarly communication between the two countries, but demonstrates that the communication is chiefly in the technological field and largely one-way at that. What really are the possibilities for fruitful communication between the capitalist republican United States and the People's Republic of China (defined in its 1975 constitution as "a dictatorship of the proletariat") in the social sciences and the humanities? In these areas can the two nations find a common language?

In conclusion let me expand upon the triangle sketched by Professor Iriye; the equation also includes the USSR. History records the Chinese/Manchu drive into Central Asia, the American drive into the West Pacific, the Japanese drive onto the Asian mainland, and also a Russian drive eastwards to the Pacific. The Pacific Doctrine enunciated by President Ford in December, 1975, took due note of the Russian presence in Northeast Asia. He remarked that "The United States, the Soviet Union, China, and Japan are all Pacific powers." He said also that "The partnership with Japan is a pillar of our strategy." He made another point of significance to our discussion, namely that "Peace in Asia requires a structure of economic cooperation reflecting the aspirations of all the peoples of the region." We are back to John Stuart Mill and List.

But if the United States has a Pacific Doctrine, so too undoubtedly have the USSR, Japan, and the People's Republic of China, and those several doctrines are not necessarily mutually compatible. In fact, as defined in the recent past by both Mao Tse-tung and Chou En-lai, the American and Chinese strategic aims are mutually antagonistic. In that quadrilateral politico-economic relationship, therefore, there is bound to be in the future as in the past substantial conflict of national objectives, and a close, easy relationship between the United States and the People's Republic of China in particular is going to be difficult to achieve: it will require the reconciliation of a number of opposites—or, shall we say, using a Maoism, of "contradictions"? But pragmatism on both sides could show the way.

O. EDMUND CLUBB
Former Director, Office of Chinese Affairs, Department of State